AF598273

THE 1785 ABSTRACT OF JAMES HUTTON'S THEORY OF THE EARTH

THE 1785 ABSTRACT OF JAMES HUTTON'S THEORY OF THE EARTH

Introduction

G. Y. Craig

Published for
Edinburgh University Library
Edinburgh University Department of Geology
and
Edinburgh Geological Society
by

SCOTTISH ACADEMIC PRESS
EDINBURGH

Published by
Scottish Academic Press Ltd,
33 Montgomery Street
Edinburgh EH7 5JX

SBN 7073 0529 2

Printed by Athenaeum Press
Newcastle upon Tyne

CONTENTS

INTRODUCTION

James Hutton's place in the history of geology is secure. Born in Edinburgh in 1726, son of a one-time city treasurer, he attended the High School and the University of Edinburgh, before taking the degree of M.D. at the University of Leyden in 1749. He farmed in the Scottish Borders for a number of years before returning to Edinburgh around 1767. During that period he became interested in geology, one of a number of subjects which he pursued with vigour in the later part of his life. His *Theory of the Earth* is undoubtedly the most original and most notable of his diverse publications. The basic theme,[1] iterated and expanded through three subsequent versions[2,3,4] is that the continents are repeatedly being wasted away and renewed by the operation of earth processes which can be seen at work at the present day, and which operated in similar fashion in the geological past. The decay of the land, its transport to the sea, the uplift and subsidence of continents, and volcanic activity are visible manifestations of the Earth operating as a machine fired by heat. The operation of cycles of decay and renewal took place over an indefinite space of time 'so that, with respect to human observation, this world has neither a beginning nor an end'.[1, p.28]

Although Hutton had formed the main elements of his theory almost 20 years previously, he had told it to only a few of his close friends. Not until 1785 did he communicate his theory to a wider audience of Fellows of the Physical Class of the Royal Society of Edinburgh. In the first volume of the *Transactions*[2] it is recorded on page 36 that on March 7, 1785 'Dr Black, in the absence of the author, read the first part of Dr Hutton's *Theory of the Earth*, which is published in this volume'. On April 4th of that year Dr Hutton was present to read the second part of his paper, which was subsequently published in 1788.[2] A much extended version appeared in two volumes in 1795[3] and an incomplete third volume was published in 1899,[4] but not until 1947 was it realized by Dr Victor Eyles that a printed Abstract of the 1785 address existed. Although the pamphlet is anonymous, there can be no doubt that Hutton was the author and that it is the first published account of his remarkable theory.[5]

I know of seven copies held in the public domain. The first[6] bears the inscription 'from the Author, Dr Hutton' and is part of the Eyles Collection gifted to the library of the University of Bristol by Joan Eyles, shortly before her death. Other places where a copy is held include the Universities of Harvard, Illinois and Virginia, the U.S. National Library of Medicine; and in Edinburgh, the Royal Botanic Garden and the Library of the University of Edinburgh. The pamphlet was reprinted by the Royal Society of Edinburgh in 1950[5] and became more widely available in North America in 1970 as part of a facsimile reproduction[7] of some of Hutton's geological papers. In 1979, a rare original copy was sold at Christie's in London for £3200.

This is the first time that a facsimile of the Abstract has been published as an entity, befittingly in the year in which the Royal Society of Edinburgh is celebrating the bicentenary of Hutton's work. In the eighteenth century the Society met in the Library of the University of Edinburgh[8, p.6] and so it is particularly appropriate that the Library's copy has been used for this reproduction. The title page has a signature in the name of 'W. Thomson' an early benefactor to the University,[9] the stamp of the library and a cataloguer's erroneous attribution of the anonymous pamphlet to Charles (correctly emended in pencil to James) Hutton, presumably the one-time Professor of Mathematics at the Royal Military Academy at Woolwich, who was a contemporary (1737–1823). The Abstract is one of the more clearly written of James Hutton's works. It contains the main elements of his Theory, but must have been printed before Hutton[10,11] had seen outcrops of granite at Glen Tilt, Blair Atholl, in 1785 and discovered in 1787 and 1788 the geological unconformities which were to provide him with such spectacular proof of the decay and renewal of continents.

It is with considerable pleasure that the Library of the University of Edinburgh, the Department of Geology and the Edinburgh Geological Society join the Scottish Academic Press in the publication of this facsimile.

All three institutions are indebted to Dr Douglas Grant, the publisher, for ensuring that the quality of the production is in keeping with the significance of the occasion.

[1] This Abstract.

[2] Hutton, J., 1788. Theory of the Earth. *Transactions of the Royal Society of Edinburgh*, 1, 209.

[3] Hutton, J., 1795. *Theory of the Earth with Proofs and Illustrations.* I and II. Cadell and Davies; London, and Creech; Edinburgh.

[4] Hutton, J., 1899. *Theory of the Earth with Proofs and Illustrations.* III. *ed.* A. Geikie, Geological Society of London.

[5] Eyles, V. A., 1950. Note on the original publication of Hutton's Theory of the Earth, and on the subsequent forms in which it was issued. *Proceedings of the Royal Society of Edinburgh*, 63B, 377.

[6] Eyles, V. A., 1955. A bibliographic note on the earliest printed version of James Hutton's Theory of the Earth, its form and date of publication. *Journal of the Society for the Bibliography of Natural History*, 3, 105.

[7] White, G. W. (*ed.*), 1970. *Contributions to the History of Geology, volume 5. James Hutton's System of the Earth. . . .* Hafner Press, New York.

[8] Campbell, N. and Smellie, R. M. S. 1983. *The Royal Society of Edinburgh (1783–1983)*. Royal Society of Edinburgh, Edinburgh.

[9] Waterston, C. D. 1965. William Thomson (1761–1806) a forgotten benefactor. *University of Edinburgh Journal*, 22, 122.

[10] Hutton, J. 1794. Observations on granite. *Transactions of the Royal Society of Edinburgh*, 3, 77.

[11] Craig, G. Y., McIntyre, D. B. and Waterston, C. D. 1977. *James Hutton's Theory of the Earth—The Lost Drawings*. Scottish Academic Press, Edinburgh.

G. Y. Craig,
Department of Geology,
University of Edinburgh.

June, 1987

ABSTRACT
OF A
DISSERTATION

The handwritten addition on page 7 states
“congelation is + [more] appropriate and exclusive”.

ABSTRACT

OF A

DISSERTATION

READ IN THE

ROYAL SOCIETY OF EDINBURGH,

UPON THE

SEVENTH OF MARCH, AND FOURTH OF APRIL,

M,DCC,LXXXV,

CONCERNING THE

SYSTEM OF THE EARTH, ITS DURATION, AND STABILITY.

By Chas Hutton M.D.

ABSTRACT, &c.

THE purpose of this Differtation is to form fome eftimate with regard to the time the globe of this Earth has exifted, as a world maintaining plants and animals; to reafon with regard to the changes which the earth has undergone; and to fee how far an end or termination to this fyftem of things may be perceived, from the confideration of that

that which has already come to paſs.

As it is not in human record, but in natural hiſtory, that we are to look for the means of aſcertaining what has already been, it is here propoſed to examine the appearances of the earth, in order to be informed of operations which have been tranſacted in time paſt. It is thus that, from principles of natural philoſophy, we may arrive at ſome knowledge of order and ſyſtem in the oeconomy of this globe, and may form a rational opinion with regard to the courſe of nature, or to events which are in time to happen.

The

The ſolid parts of the preſent land appear, in general, to have been compoſed of the productions of the ſea, and of other materials ſimilar to thoſe now found upon the ſhores. Hence we find reaſon to conclude,

1*ſt*, That the land on which we reſt is not ſimple and original, but that it is a compoſition, and had been formed by the operation of ſecond cauſes.

2*dly*, That, before the preſent land was made, there had ſubſiſted a world compoſed of ſea and land, in which were tides and currents, with ſuch operations at the

the bottom of the ſea as now take place. And,

Laſtly, That, while the preſent land was forming at the bottom of the ocean, the former land maintained plants and animals; at leaſt, the ſea was then inhabited by animals, in a ſimilar manner as it is at preſent.

Hence we are led to conclude, that the greater part of our land, if not the whole, had been produced by operations natural to this globe; but that, in order to make this land a permanent body, reſiſting the operations of the waters, two things had been requi-

red;

red ; *1ſt*, The * conſolidation of maſſes formed by collections of looſe or incoherent materials ; *2dly*, The elevation of thoſe conſolidated maſſes from the bottom of the ſea, the place where they were collected, to the ſtations in which they now remain above the level of the ocean.

Here are two different changes, which may ſerve mutually to throw

* There are two ſenſes in which the term *ſolidity* is uſed ; one of theſe is in oppoſition to *fluidity*, the other to *vacuity*. When the change from a fluid ſtate to that of ſolidity, in the firſt ſenſe, is to be expreſſed, we ſhall employ the term *concretion* ;✗ conſequently, the conſolidation of a maſs is only to be underſtood as in oppoſition to its vacuity, or porouſneſs. ✗Congelation is the appropriate, & exclusive—

throw ſome light upon each other; for, as the ſame ſubject has been made to undergo both theſe changes, and as it is from the examination of this ſubject that we are to learn the nature of thoſe events, the knowledge of the one may lead us to ſome underſtanding of the other.

Thus the ſubject is conſidered as naturally divided into two branches, to be ſeparately examined: *Firſt*, by what natural operation ſtrata of looſe materials had been formed into ſolid maſſes; *ſecondly*, By what power of nature the conſolidated ſtrata at the

the bottom of the ſea had been transformed into land.

With regard to the *firſt* of theſe, the conſolidation of ſtrata, there are two ways in which this operation may be conceived to have been performed; firſt, by means of the ſolution of bodies in water, and the after concretion of theſe diſſolved ſubſtances, when ſeparated from their ſolvent; *ſecondly*, the fuſion of bodies by means of heat, and the ſubſequent congelation of thoſe conſolidating ſubſtances.

With regard to the operation of water, it is *firſt* conſidered, how far the power of this ſolvent, acting in the natural ſituation of thoſe ſtrata, might be ſufficient to produce the effect; and here it is found, that water alone, without any other agent, cannot be ſuppoſed capable of inducing ſolidity among the materials of ſtrata in that ſituation. It is, *2dly*, conſidered, how far, ſuppoſing water capable of conſolidating the ſtrata in that ſituation, it might be concluded, from examining natural appearances, that this had been actually the caſe? Here again, having proceeded upon this principle, that water could only conſolidate ſtrata with ſuch ſubſtances as it has the power to diſſolve,

and

and having found ſtrata conſolidated with every ſpecies of ſubſtance, it is concluded, that ſtrata in general have not been conſolidated by means of aqueous ſolution.

With regard to the other probable means, heat and fuſion, theſe are found to be perfectly competent for producing the end in view, as every kind of ſubſtance may by heat be rendered ſoft, or brought into fuſion, and as ſtrata are actually found conſolidated with every different ſpecies of ſubſtance.

A more particular discussion is then entered into: Here, consolidating substances are considered as being classed under two different heads, viz. Siliceous and sulphureous bodies, with a view to prove, that it could not be by means of aqueous solution that strata had been consolidated with those particular substances, but that their consolidation had been accomplished by means of heat and fusion.

Sal Gem, as a substance soluble in water, is next considered, in order to show that this body had been last in a melted state; and this example is confirmed by one of of fossile alkali. The case of particular septaria of iron-stone, as well as certain crystallized cavities in mineral bodies, are then given as examples of a similar fact; and as containing, in themselves, a demonstration, that all the various mineral substances had been concreted and crystallized immediately from a state of fusion.

Having thus proved the actual fusion of the substances with which strata had been consolidated, in having such fluid bodies introduced among their interstices, the case of strata, consolidated by means of the simple fusion of their proper materials, is next considered;

ed; and examples are taken from the moſt general ſtrata of the globe, viz. ſiliceous and calcareous. Here alſo demonſtration is given, that this conſolidating operation had been performed by means of fuſion.

Having come to this general concluſion, that heat and fuſion, not aqueous ſolution, had preceded the conſolidation of the looſe materials collected at the bottom of the ſea, thoſe conſolidated ſtrata, in general, are next examined, in order to diſcover other appearances, by which the doctrine may be either confirmed or refuted. Here the changes of ſtrata, from their

their natural ſtate of continuity, by veins and ſiſſures, are conſidered; and the cleareſt evidence is hence deduced, that the ſtrata have been conſolidated by means of fuſion, and not by aqueous ſolution; for, not only are ſtrata in general found interſected with veins and cutters, an appearance inconſiſtent with their having been conſolidated ſimply by previous ſolution; but, in proportion as ſtrata are more or leſs conſolidated, they are found with the proper correſponding appearances of veins and fiſſures.

With regard to the ſecond branch, in conſidering by what power

power the consolidated strata had been transformed into land, or raised above the level of the sea, it is supposed that the same power of extreme heat, by which every different mineral substance had been brought into a melted state, might be capable of producing an expansive force, sufficient for elevating the land, from the bottom of the ocean, to the place it now occupies above the surface of the sea. Here we are again referred to nature, in examining how far the strata, formed by successive sediments or accumulations deposited at the bottom of the sea, are to be found in that regular state, which would necessarily take place

in

in their original production; or if, on the other hand, they are actually changed in their natural situation, broken, twisted, and confounded, as might be expected, from the operation of subterranean heat, and violent expansion. But, as strata are actually found in every degree of fracture, flexure, and contortion, consistent with this supposition, and with no other, we are led to conclude, that our land had been raised above the surface of the sea, in order to become a habitable world; as well as that it had been consolidated by means of the same power of subterranean heat, in order to remain above the level of the sea,

 and

and to reſiſt the violent efforts of the ocean.

This theory is next confirmed by the examination of mineral veins, thoſe great fiſſures of the earth, which contain matter perfectly foreign to the ſtrata they traverſe; matter evidently derived from the mineral region, that is, from the place where the active power of fire, and the expanſive force of heat, reſide.

Such being conſidered as the operations of the mineral region, we are hence directed to look for the manifeſtation of this power and force, in the appearances of

nature.

nature. It is here we find eruptions of ignited matter from the ſcattered volcano's of the globe; and theſe we conclude to be the effects of ſuch a power preciſely as that about which we now inquire. Volcano's are thus conſidered as the proper diſcharges of a ſuperfluous or redundant power; not as things accidental in the courſe of nature, but as uſeful for the ſafety of mankind, and as forming a natural ingredient in the conſtitution of the globe.

The doctrine is then confirmed, by examining this earth, and by finding every where, beſide the many marks of ancient volcano's, cano's, abundance of ſubterraneous or unerupted lava, in the baſaltic rocks, the Swediſh trap, the toadſtone, ragſtone, and whinſtone of Britain and Ireland, of which particular examples are cited; and a deſcription given of the three different ſhapes in which that unerupted lava is found.

The peculiar nature of this ſubterraneous lava is then examined; and a clear diſtinction is formed between this baſaltic rock and the common volcanic lavas.

Laſtly, The extenſion of this theory, reſpecting mineral ſtrata, to all parts of the globe, is made, by by finding a perfect ſimilarity in the ſolid land through all the earth, although, in particular places, it is attended with peculiar productions, with which the preſent inquiry is not concerned.

A theory is thus formed, with regard to a mineral ſyſtem. In this ſyſtem, hard and ſolid bodies are to be formed from ſoft bodies, from looſe or incoherent materials, collected together at the bottom of the ſea; and the bottom of the ocean is to be made to change its place with relation to the centre of the earth, to be formed into land above the level

of

of the ſea, and to become a country fertile and inhabited.

That there is nothing viſionary in this theory, appears from its having been rationally deduced from natural events, from things which have already happened; things which have left, in the particular conſtitutions of bodies, proper traces of the manner of their production; and things which may be examined with all the accuracy, or reaſoned upon with all the light, that ſcience can afford. As it is only by employing ſcience in this manner, that philoſophy enlightens man with the knowledge of that wiſdom or

deſign

deſign which is to be found in nature, the ſyſtem now propoſed, from unqueſtionable principles, will claim the attention of ſcientific men, and may be admitted in our ſpeculations with regard to the works of nature, notwithſtanding many ſteps in the progreſs may remain unknown.

By thus proceeding upon inveſtigated principles, we are led to conclude, that, if this part of the earth which we now inhabit had been produced, in the courſe of time, from the materials of a former earth, we ſhould, in the examination of our land, find data from which to reaſon, with re-

gard

gard to the nature of that world, which had exifted during the period of time in which the prefent earth was forming; and thus we might be brought to underftand the nature of that earth which had preceded this; how far it had been fimilar to the prefent, in producing plants and nourifhing animals. But this interefting point is perfectly afcertained, by finding abundance of every manner of vegetable production, as well as the feveral fpecies of marine bodies, in the ftrata of our earth.

Having thus afcertained a regular fyftem, in which the prefent land of the globe had been firft

formed

formed at the bottom of the ocean, and then raiſed above the ſurface of the ſea, a queſtion naturally occurs with regard to time; what had been the ſpace of time neceſſary for accompliſhing this great work?

In order to form a judgment concerning this ſubject, our attention is directed to another progreſs in the ſyſtem of the globe, namely, the deſtruction of the land which had preceded that on which we dwell. Now, for this purpoſe, we have the actual decay of the preſent land, a thing conſtantly tranſacting in our view, by which to form an eſtimate.

This decay is the gradual ablution of our ſoil, by the floods of rain; and the attrition of the ſhores, by the agitation of the waves.

If we could meaſure the progreſs of the preſent land, towards its diſſolution by attrition, and its ſubmerſion in the ocean, we might diſcover the actual duration of a former earth; an earth which had ſupported plants and animals, and had ſupplied the ocean with thoſe materials which the conſtruction of the preſent earth required; conſequently, we ſhould have the meaſure of a correſponding ſpace of time, viz. that which had been required in the production of the preſent

present land. If, on the contrary, no period can be fixed for the duration or destruction of the present earth, from our observations of those natural operations, which, though unmeasurable, admit of no dubiety, we shall be warranted in drawing the following conclusions; 1*st*, That it had required an indefinite space of time to have produced the land which now appears; 2*dly*, That an equal space had been employed upon the construction of that former land from whence the materials of the present came; *lastly*, That there is presently laying at the bottom of the ocean the foundation of future land, which is to

appear

appear after an indefinite ſpace of time.

But, as there is not in human obſervation proper means for meaſuring the waſte of land upon the globe, it is hence inferred, that we cannot eſtimate the duration of what we ſee at preſent, nor calculate the period at which it had begun; ſo that, with reſpect to human obſervation, this world has neither a beginning nor an end.

An endeavour is then made to ſupport the theory by an argument of a moral nature, drawn from the conſideration of a final cauſe. Here a compariſon is form-

ed

ed between the preſent theory, and thoſe by which there is neceſſarily implied either evil or diſorder in natural things; and an argument is formed, upon the ſuppoſed wiſdom of nature, for the juſtneſs of a theory in which perfect order is to be perceived. For,

According to the theory, a ſoil, adapted to the growth of plants, is neceſſarily prepared, and carefully preſerved; and, in the neceſſary waſte of land which is inhabited, the foundation is laid for future continents, in order to ſupport the ſyſtem of this living world.

Thus,

Thus, either in ſuppoſing Nature wiſe and good, an argument is formed in confirmation of the theory, or, in ſuppoſing the theory to be juſt, an argument may be eſtabliſhed for wiſdom and benevolence to be perceived in nature. In this manner, there is opened to our view a ſubject intereſting to man who thinks; a ſubject on which to reaſon with relation to the ſyſtem of nature; and one which may afford the human mind both information and entertainment.